STARTING OUT

FREIGHT TRAINS

MEG GREVE

CREATIVE EDUCATION • CREATIVE PAPERBACKS

CONT

ENTS

I SEE A FREIGHT TRAIN.

I wait at a railroad crossing.

The gate is down.

I hear the train coming down the tracks.

ZIM
EVERGREEN
OOCL
8854

The engine, or locomotive, is powerful. It pulls a long line of cars.

Flat cars carry big containers. After the train stops, the containers can go onto trucks.

A freight train moves cargo.

This train is long!

It has many kinds of cars.

An open hopper car can carry coal. A tank car can carry gasoline. A gondola can carry logs.

MAKE A NOISE

WHO

oooo!

Can you make the sounds of a freight train?

Listen to these sounds:

https://www.youtube.com/watch?v=ae8XeWzzw-Y

Now it is your turn!

FREIGHT TRAIN WORDS

cargo: the things freight trains move from place to place

coal: a rock that burns easily and that provides energy for making things work

gasoline: a liquid fuel that powers cars and other vehicles

tracks: the metal rails that a train travels on

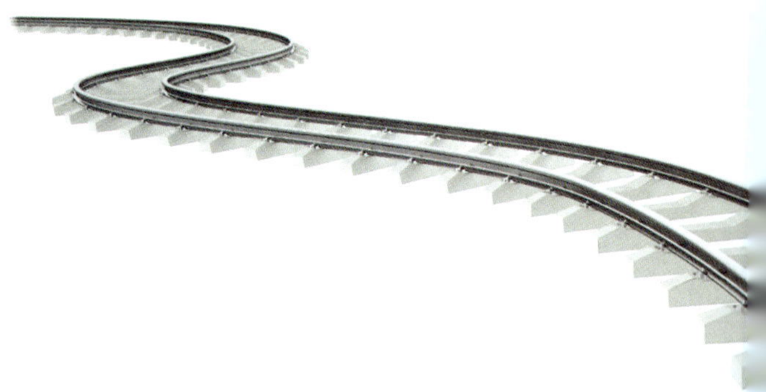

READING CORNER

Gleisner, Jenna Lee. *Freight Trains*. Minneapolis, Minn.: Jump!, 2020.

Prenat, Sophie. *Ultimate Spotlight: Trains.* Miami, Fla.: Twirl, 2020.

Shealy, Dennis R. *My Little Golden Book About Trains.* New York: Little Golden Books, 2021.

INDEX

PUBLISHED BY CREATIVE EDUCATION AND CREATIVE PAPERBACKS
P.O. Box 227, Mankato, Minnesota 56002
Creative Education and Creative Paperbacks are imprints of The Creative Company
www.thecreativecompany.us

LIBRARY OF CONGRESS CATALOGING-IN-PUBLICATION DATA
Names: Greve, Meg, author.
Title: Freight trains / by Meg Greve.
Description: Mankato, Minnesota : Creative Education and Creative Paperbacks, [2025] | Series: Starting out | Includes bibliographical references and index. | Audience: Ages 4-7 | Audience: Grades K-1 | Summary: "Freight trains will introduce budding book learners to a noisy, colorful world with this new Starting Out title. Colorful photos, labeled diagrams, 'Make a Noise' section, glossary, and more ignite a passion for learning"-- Provided by publisher.
Identifiers: LCCN 2023059445 (print) | LCCN 2023059446 (ebook) | ISBN 9798889891734 (library binding) | ISBN 9781682775585 (paperback) | ISBN 9798889891857 (ebook)
Subjects: LCSH: Railroad trains--Juvenile literature. | Freight cars--Juvenile literature. | CYAC: Railroad trains. | Freight cars. | LCGFT: Instructional and educational works.
Classification: LCC TF148 .G745 2025 (print) | LCC TF148 (ebook) | DDC 625.2--dc23/eng/20240130
LC record available at https://lccn.loc.gov/2023059445
LC ebook record available at https://lccn.loc.gov/2023059446

DESIGN AND PRODUCTION
Design by Rhea Magaro and Jennifer Bowers
Production by Beeline Media and Design
Art direction by Tom Morgan
Printed in the United States of America

PHOTOGRAPHS by AdobeStock (Maharketing), Alamy (Robert McGouey / Industry), Shutterstock (3DMI, freestore 839, Nerthuz, Kenneth Sponsler, Ton Bangkeaw, Ian Dewar Photography, Maksim Safaniuk, Vadi Fuoco, kemdim, jeffy11390, Bohbeh, Rost9)